SHOE

FACTORY

EFFICIENCY

A History of Shoemaking

Shoemaking, at its simplest, is the process of making footwear. Whilst the art has now been largely superseded by mass-volume industrial production, for most of history, making shoes was an individual, artisanal affair. 'Shoemakers' or 'cordwainers' (cobblers being those who repair shoes) produce a range of footwear items, including shoes, boots, sandals, clogs and moccasins – from a vast array of materials.

When people started wearing shoes, there were only three main types: open sandals, covered sandals and clog-like footwear. The most basic foot protection, used since ancient times in the Mediterranean area, was the sandal, which consisted of a protective sole, attached to the foot with leather thongs. Similar footwear worn in the Far East was made from plaited grass or palm fronds. In climates that required a full foot covering, a single piece of untanned hide was laced with a thong, providing full protection for the foot, thus forming a complete covering. These were the main two types of footwear, produced all over the globe. The production of wooden shoes was mainly limited to medieval Europe however – made from a single piece of wood, roughly shaped to fit the foot.

A variant of this early European shoe was the clog, which were wooden soles to which a leather upper was attached. The sole and heel were generally made from one piece of maple or ash two inches thick, and a little longer and broader than the desired size of shoe. The outer side of the sole and heel was fashioned with a long chisel-edged implement, called the clogger's knife or stock; while a second implement, called the groover, made a groove around the side of the sole. With the use of a 'hollower', the inner sole's contours were adapted to the shape of the foot. In even colder climates, such designs were adapted with furs wrapped around the feet, and then sandals wrapped over them. The Romans used such footwear to great effect whilst fighting in Northern Europe, and the native Indians developed similar variants with their ubiquitous moccasin.

By the 1600s, leather shoes came in two main types. 'Turn shoes' consisted of one thin flexible sole, which was sewed to the upper while outside in and turned over when completed. This type was used for making slippers and similar shoes. The second type united the upper with an insole, which was subsequently attached to an out-sole with a raised heel. This was the main variety, and was used for most footwear, including standard shoes and riding boots.

Shoemaking became more commercialized in the mid-eighteenth century, as it expanded as a cottage industry. Large warehouses began to stock footwear made by many small manufacturers from the area. Until the nineteenth century, shoemaking was largely a traditional handicraft, but by the century's end, the process had been almost completely mechanized, with production occurring in large factories. Despite the obvious economic gains of mass-production, the factory system produced shoes without the individual differentiation that the traditional shoemaker was able to provide.

The first steps towards mechanisation were taken during the Napoleonic Wars by the English engineer, Marc Brunel. He developed machinery for the mass-production of boots for the soldiers of the British Army. In 1812 he devised a scheme for making nailed-boot-making machinery that automatically fastened soles to uppers by means of metallic pins or nails. With the support of the Duke of York, the shoes were manufactured, and, due to their strength, cheapness, and durability, were introduced for the use of the army. In the same year, the use of screws and staples was patented by Richard Woodman. However, when the war ended in 1815, manual labour became much cheaper again, and the demand for military equipment subsided. As a consequence, Brunel's system was no longer profitable and it soon ceased business.

Similar exigencies at the time of the Crimean War stimulated a renewed interest in methods of mechanization and mass-production, which proved longer lasting. A shoemaker in Leicester, Tomas Crick, patented the design for a riveting machine in 1853. He also introduced the use of steam-powered rolling-machines for hardening leather and cutting-machines, in the mid-1850s. Another important factor in shoemaking's mechanization, was the introduction of the sewing machine in 1846 – a development which revolutionised so many aspects of clothes, footwear and domestic production.

By the late 1850s, the industry was beginning to shift towards the modern factory, mainly in the US and areas of England. A shoe stitching machine was invented by the American Lyman Blake in 1856 and perfected by 1864.

Entering in to partnership with Gordon McKay, his device became known as the McKay stitching machine and was quickly adopted by manufacturers throughout New England. As bottlenecks opened up in the production line due to these innovations, more and more of the manufacturing stages, such as pegging and finishing, became automated. By the 1890s, the process of mechanisation was largely complete.

Traditional shoemakers still exist today, especially in poorer parts of the world, and do continue to create custom shoes. In more economically developed countries however, it is a dying craft. Despite this, the shoemaking profession makes a number of appearances in popular culture, such as in stories about shoemaker's elves (written by the Brothers Grimm in 1806), and the old proverb that 'the shoemaker's children go barefoot.' Chefs and cooks sometimes use the term 'shoemaker' as an insult to others who have prepared sub-standard food, possibly by overcooking, implying that the chef in question has made his or her food as tough as shoe leather or hard leather shoe soles. Similarly, reflecting the trade's humble beginnings, to 'cobble' can mean not only to make or mend shoes, but 'to put together clumsily; or, to bungle.'

As is evident from this short introduction, 'shoemaking' has a long and varied history, starting from a simple means of providing basic respite from the elements, to a fully mechanised and modern, global trade. It is able to provide a fascinating insight not only into fashion, but society, culture and climate more generally. We hope the reader enjoys this book.

SHOE FACTORY EFFICIENCY

By

John E. Kirwin

CONTENTS

Chapter I. - Unit Area of Production
" II. - Labor
" III. - Machinery
" IV. - The Productive System
" V. - Specialization
" VI. - Departments
" VII. - Shoe Costs
" VIII. - Conclusion

CHAPTER I.

Unit Area of Production

Shoe manufacturers sometimes are at a loss to know whether or not they are making the best possible use of the floor space at their disposal. Careful investigations are made and the untiring efforts of the superintendent are enlisted. Frequently the concerns in doubt go outside and employ the services of men whose business it is to study, devise and improve systems with an eye to greater productiveness. This latter course is a very profitable investment, providing the men engaged are thoroughly conversant with the pecular problem of producing shoes.

It is very natural that shoe manufacturers should want to know exactly where they are at, inasmuch as the field of competition is extremely keen and the margin of profit small. The race is, indeed, for the sure-footed no less than for the fleet-footed. A few suggestions may clear the situation. For instance, a manufacturer may perceive in advance that the price of materials is going to rise and in this case he is confronted with the problem of "stocking up". Now, unless he knows down to a cent what it costs him to make his product and unless he is informed down to a square foot as to how much room it requires to move the work through to completion, he cannot wisely venture into the market and place heavy orders with a feeling of clear confidence. Lacking such precise knowledge of his own business he falls into the habit of waiting and considering, with the result later of buying materials at advanced cost. Many manufacturers take long chances and they stand to win or lose, but those best equipped with a precise conception of their productive capacity and the cost per unit of product are the men who seize opportunities at the right time and then push their competitors. It is well known that some men are classed as conservative who are prevented from acting because of a lack of complete understanding of their own working organization. Others are stigmatized as adventurers, who merely act quickly and decisively because of well placed confidence in the efficiency of their productive power.

The point which is being driven home is that too few manufacturers know within a row of apple trees just how much floor space it takes to turn out a unit of their product, which is a pair of shoes. This is one of the units of productive efficiency which should be ascertained exactly.

The point just made involves a good deal. It comprehends all the parts of the factory, department by department, and their correlation. Much depends on the physical structure of the factory, in attaining the greatest departmental efficiency, or in other words an arrangement which might be admirably adapted to one shoe manufacturing plant would be the worst possible arrangement in another. It is, therefore, necessary for each plant to line up its forces to the highest productive advantage. Some observations on how certain factories are arranged may clarify these claims.

More and more the belief is gaining ground that the cutting of the upper leather and sole leather should not be a working part of the factory proper. Hence we note that some large firms are segregating the cutting unit from the organization proper. At present this arrangement seems to apply chiefly to concerns engaged on a large scale. The whole idea back of it is of course, that so much money may be either made or lost in the cutting up of the leather that it should be made a seperate department from all the rest of the factory. In other words, the leather should be received checked, sorted, and cut, and then be sold to the factory proper, just as though it were a purchase made outside. Needless to say, an unmistakable double check is hereby afforded on the most expensive material used in the process of shoe manufacturing.

In some factories that are built suitably it is found convenient to have the cutting and fitting departments on the same floor. Wherever this is in vogue it is noted that many of the operations usually done in the stitching room are transferred to the cutting room. Such operations as skiving, cementing, folding, lining and upper marking—which are preliminary factors in the work—and are found to facilitate greatly the fast rotating of the shoes through the stitching room, if they are done in the cutting room. Investigations show that in many instances, where shoes are held up in the stitching room, the delay is to be traced to one or the other of the preliminary operations. When work is once well started in this depart ment there is little fear of it being held up.

To carry the observations further, it is found that greater efficiency is often gained by having the making and the finishing departments on the same floor. Where such an arrangement obtains, poor edges or poor work of any sort may be pushed back quickly to the department from which it just issued, without, it must be noted, the necessity of transferring on the elevator. Applying the same principle, an arrangement which provides for the lasting and stock-fitting departments on the same floor may be defended on the grounds of greater productive efficiency.

It is not deemed practical to go too much into specific details as this discussion is merely intended to furnish food for thought and to call attention to the fact that shoe manufacturers are in a position which demands the acquisition of every bit of detailed knowledge relating to their individual plants which it is possible for them to secure. The reason for this is, as we have pointed out, because the competition is becoming so great and because the net profits are being shaved so closely. It should be further stated that small factories are undeniably on the decrease, and there must be a good reason back of it.

CHAPTER II.

Labor Conditions Must Be Right.

Much has already been left unsaid, but as it is my present intention merely to break the ground, the gamut of details, such as would be necessary in a fine-tooth analysis, is purposely limited. The fact that hardly any two factories in the country produce shoes under identically the same conditions is well established. The further fact that hardly any two factories sufficiently alike in the amount and quality of their products to admit of comparison use the same amount of labor or attain the same labor efficiency is a subject that invites close and thoughtful attention.

Geographical situation admittedly has a great effect. Sectional location within that same geographical unit has likewise a great effect. These are conditions that may be taken advantage of, but cannot well be changed. The main point at issue, however, is the efficient production of labor. Conditions, let me hasten to say, have much to do with this, and that they must be right in order to attain the desired end scarcely needs to be stated. Two sets of conditions have been suggested, but the latter, or internal conditions, are the ones that are now to be considered.

The construction and physical arrangement of the factory helps or hinders the best production. Room enough for each and every operator at the machine or at the bench is hardly less important than the proper lighting, heating and ventilating of the various work rooms. Those who have seen cutters at the bench swinging their arms, bundling themselves in wearing apparel and blowing on their fingers, know the effect of such conditions as regards production. Poorly ventilated stitching rooms with girls fainting every now and then is a condition not to be overlooked. Misunderstandings and quarrels between operatives who have been unable to keep their work carefully seperated because of cramped quarters also suggests wrong conditions. See that the operatives have room, light, heat and the best air that can be furnished and the productive efficiency will not be hindered on this score.

The piece-price system is undoubtedly a good producer in itself, but even this system is often greatly misunderstood. The fact that the firm pays each operative on this basis just what the operative earns should not be the end and aim in itself as is often the case. Not a few operatives in every factory are satisfied with a little less than they might earn if conditions were slightly better. This indeterminable margin is really a dead loss to the concern, although it might tax the persuasive powers of a great orator to make some manufacturers believe it. The fact remains, however, that it costs the manufacturer no more for power and running expense to get this additional output per piece-price operative than it does not to get it. There is a deep-rooted belief among manufacturers that the danger flag only flies over the heads of day laborers and that they alone require strict tabs to be kept. With the more progressive concerns and those who analyze their expense more closely, the fact that piece-price operatives will take care of themselves is not given this credence.

The workmen from the first to the last operation in the shoe factory require careful and painstaking supervision. Here is an important condition. It should be said, and experience attests the fact, that quantity of supervision will not solve the problem. It must be productive supervision or it is worse than none. Each factory must depend on itself in this matter, but some record should be kept to show that the supervision is highly productive. Many foremen who have been good producers at the machine have proved absolutely inadequate to the task of getting production out of others. And this applies equally to positions higher up in the organization.

There is another important matter that demands careful attention. Speaking largely, the laborers fall into two distinct classes, the skilled and the unskilled. How many concerns take this into careful consideration and place craftsmanship at a premium? There is no possible denial of the fact that a factory which is so organized that a skilled laborer does nothing but the work that requires skill, with the unskilled laborers supplementing him, is far in advance of other factories not discriminating thus carefully. Such discrimination is absolutely essential to the highest factory efficiency, and no point mentioned thus far challenges closer thought on the part of the reader. The writer has known instances where skilled operators have spent from thirty to fifty per cent. of their time each day on operations that could not possibly fall in the group of expert labor. No doubt the differentiation along this line obtained far more widely when the system of apprenticeship was in vogue. It is not necessary, however, to return to the old system of apprentices to make this careful division of labor. The principle itself is operative, furthermore, from the bottom to the top of the factory organization.

The fact that many things may be accomplished under adverse conditions is no argument whatever. It is rather an advocacy of the principle that right conditions make for greater efficiency. But better proof than this is at hand. Factories are being run to-day which take careful account of such elements as are mentioned above. Practical working demonstrations carry their own conviction. Do not fail to analyze the labor account in your efforts to promote the productive efficiency of your factory.

CHAPTER III.

Machinery

I have discussed one of the chief productive units, namely, labor. The fact might have been stated that, although labor is a vital factor and the one which can be least meddled with, it is the first object of attack when a shoe manufacturer readjusts or institutes a change in his expense accounts. I tried to show that conditions must be right to get the highest labor efficiency, and in the following discussion on machinery, the other great productive unit, right conditions must not be lost to view.

A machine is invented and put on the market to accomplish certain definite results. From time immemorial machines have been supplementing the workman and accomplishing different phases of the work that previously had been done by hand. The theory that machines throw men out of employment has been exploded time and again. The fact that this theory gained early credence, which has not even now entirely passed away, is that a new machine necessarily brings about a change for a short time, but usually heralds greater production and the need of more hands. Each machine, as already stated, is made to do certain definite things under proper and specified conditions, and it will always do those things if these conditions obtain and it will continue in the service until it is worn out. A machine, however, has no brains or human intuition and will run when power is applied exactly as the arrangement of shafting, cams, pawls, rachets, levers, etc., permit it to run. Most machines will do a little more than they are intended to do, but it is not the machine in this case; it is the skill and the brain of the opperator that does it. So much about machines is agreed to by all.

Machines, not unlike a company of soldiers, should be lined up to the very best advantage. Some concerns employ a superintendent of construction. whose business it is to accomplish this end and his position is a permanent one. There is no question but that money is made and lost on the machines, the manner of handling and the general line-up. An advantage of great importance is here implied, namely, that the foreman is usually too busy with ever-present duties to reflect on the arrangement of his room with respect to the machines, but is never too busy to bring facts or suggestions to the attention of a man whose duty it is to examine and consider them. The desire for greater machine efficiency has given the position of superintendent of construction a permanent place in the shoe factory.

A separate machine ledger is kept by some shoe firms for the individual machines, and its value is unquestionable. The price of a machine, its probable life, the machine use rate, its depreciation, its intrinsic value as old junk—all these things, together with the number and position of the machine in the factory and the debit column for repairs and new parts, suggest how strictly the tabs are kept on the individual machines in certain factories.

To get the most efficient service from each and every machine it is not necessary to run them at abnormal speed; in fact, such a state of things is absolutely dead wrong. Careful tutoring among the different machine operators from the first to the last in the factory as regards the purposes, the parts, and the exact uses of each separate machine is very essential to efficient service. Careful directions as to the time of oiling, the parts to be oiled, and the cleanly condition of each machine must be known, and, most important of all, followed to the letter by every machine operator in the factory.

The unnecessarily hard strain that a machine is subjected to when slightly out of order but still running may cripple the efficiency of that machine to an extent often little realized.

Careful supervision is once more emphasized as an essential factor in gaining greater productiveness. The machine will do the certain thing that it is intended to do, providing the conditions are right. Furthermore, it will perform these specific functions far more regularly and with much less liability of being incapacitated than any human being can do. Let the correct conditions prevail and the eye of supervision be ever present and each machine will produce up to its capacity. That the machines so often are not producing up to their capacity should open the eyes of the manufacturer to the fact that wrong conditions are existing in his plant. The more progressive manufacturers are thoroughly awake to this and are reaping an advantage in increased factory efficiency.

There is no part of the shoe factory that depreciates so fast as the machinery, and no other item under repairs amounts to anything like so much as machinery repairs. Waste power is in itself an important feature, and while the machinery equipment cannot be so managed as to do away entirely with some waste power, yet this item may be kept down to a minimum. All the expense and loss that may be incurred through machinery demands that each shoe concern, which hopes to maintain its place in the advance of this great industry, must gird up its loins and fight determinedly for everything that promotes greater efficiency. This really means greater production at the same or at a less cost than previously. The productive unit of machinery should be nicely ascertained, and if it is not what it should be, then vigorous action should be taken to attain the desired end.

CHAPTER IV.
The Productive System.

With an eye to efficient factory organization many leading features have already been pointed out and progressive ideas advanced. This part of the treatment has to do with the so-called sheet system of putting work through the factory. It is applicable in any factory regardless of size, but it seems to be greatly misunderstood. In principle it means the movement of the shoes with military precision through the factory, from the cutting room to the shipping room.

There is a more or less prevailing notion that the sheet system of turning out shoes is only possible where such and such conditions obtain. Not a few manufacturers have been heard to say that they had tried out the system thoroughly and that it was not practical for their purposes. A great many times the system has failed to work properly in certain factories but this failure is no reflection on the system itself; it is a sad commentary on the conditions existing in some factories. The system itself is a great producer, but it is not one of those automatic affairs that work of their own accord. Systems of all kinds have to be developed; there is a necessary period of growth and evolution before any system can incorporate itself into the smooth working organization of the factory. Modifications which have to be made in one factory may not have to be considered at all in another. The general conditions have to be analyzed in order to institute a a system and make it fit. It is dead wrong, however, that such a productive agency should be discredited in factories where it was neither started right, nor allowed to become welded into the working organizatation of the factory.

It is not necessary to go into the details of the sheet system. In its main outlines it represents the different case numbers, styles, lasts, patterns, number of pairs, etc., which make up a day's work. Each sheet represents a day's work, and if the obstacles to production have been carefully and properly itemized, such as lasts and different styles of shoes and different kinds of stock, there should be no failure to move a sheet each day. Unless the sheets are strictly maintained and held to closely, the system can never be anything but a miserable failure. There are not a few factories that claim to be operated on the sheet system which are not in any degree managed in this manner. There is no such thing as a sheet system except in factories where the sheets produce their face value each day.

Naturally this method of getting work through the factory and shipped out on time would never have gained the prestige that it has were it not for the fact that some of the most progressive concerns in the country discovered the productive efficiency of a system whereby each department of the factory is held to a certain definite output each and every working day of the year. The genius of the system consists in the fact that obstacles and hindrances of all kind are taken into consideration in making up the sheets for each day; so that production will not be slowed down because of too many cases of button shoes or too many shoes on the same last, etc., appearing in a particular sheet. The sheets set a speed for all the factory departments which must be maintained.

One of the most common reasons why the sheet system fails to work properly in many factories is that it is allowed to follow the lines of least resistance. In other words, it may be worked successfully in the cutting and the stitching rooms, but be allowed to go astray in the lasting room. As soon as any one of the several units fails to keep up its end, then the whole system is of no avail. You cannot have a sheet system for some of the rooms and not for all. A system in operation cannot possibly be greater than any of its component parts as an affective working agency. Half-hearted or ill-advised conduct in instituting the sheet system in the shoe factory has led not a few manufacturers to believe that it is a thing that may work all right or may not, that may be a good thing and possibly a necessity in some places, but a ticklish or risky thing to meddle with. It is really a shame that a system which is such a good producer should have received so much discredit.

It is easy enough to see that a shoe firm of to-day is obliged to operate at a capacity proportionate to the general standard automatically fixed by the trade. When fewer shoes were produced, speaking generally, profits were considerably higher than at the present time, when the race is on for rapid production. Undoubtedly as the business continues, the need will more and more be felt to attain to greater efficiency. Some concerns that have tried the sheet system and failed will eventually find that it is not a question of take it or leave it, but a prime necessity to adopt it.

It is rather dangerous business to attempt to borrow and transplant systems. As has been suggested, conditions must be known and must be made right. A system needs to grow and pass through a series of evolutions. These evolutions may cost the firm money for the time being, but those who pin their faith to the sheet system will not come far amiss, and that they will thereby become greater producers is an assured fact. Time, money and patience, no less than eternal vigilance, is the price of greater productiveness.

One of the best tests that a factory is efficient is whether or not the sheet system can be successfully worked without too much difficulty. This in itself gives a correct line on the conditions prevailing in the factory. Although conditions as thus applied is a rather general term, it is thought best to use it rather than to cite concrete instances, which might easily be done. This is a system that has proved to be a remarkable producer; it has never really failed to give results; it is constantly gaining prestige; it is distinctly American; and it deserves a prominent place in all considerations along the lines of factory efficiency. Because it makes for rapid production and reduces factory difficulties and expenses, I urge it as a permanent factor in the great shoe industry.

CHAPTER V.
Specialization.

The great tendency towards specialization in the shoe manufacturing business deserves separate treatment in this discussion of factory efficiency. The term specialization is sometimes misunderstood. A firm, for instance, may have specialties or it may specialize in certain ways, but this is not specialization in its larger sense. As correctly interpreted, specialization means the careful lining up of the forces of production with the one aim of producing a certain definite article better in quality and more complete in itself than can be turned out under any other conditions. It means the thorough development and exploitation of the trade on one or more specified products.

The principle itself is thoroughly practical. A firm which makes nothing but misses' and children's shoes is in a position to locate its plant or plants in the most strategic position as regards the labor market and to develop each and all the workmen to a greater degree of perfection in this line of production than is possible for those engaged in turning out other products at the same time. Each individual workman at the bench or at the machine will gain a proficiency often little realized if he is allowed to labor at the same kind of work week in and week out. This is true from the first to the last worker in the factory, and the result is greatly increased production and better quality. Naturally a concern which caters to the wants of a particular trade is going to win that trade over and hold it against all comers who do not thus specialize.

Some concerns have even gone so far as to limit each factory to a certain kind of stock—kid shoes in one plant, patent leather shoes in another plant, calf shoes in still another plant, etc. This specialization suggests a great many possibilities, and, since leather is such a variable factor at its best, it would seem to be a natural evolution as the industry gradually becomes more solidified. As one contemplates these things it appears that the shoe business is being reclaimed from a state of chaos, which, after all, really is the case.

Every effort is being exerted by progressive shoe concerns at the present time to make their product uniform. The old-fashioned idea that certain large sizes in a case of shoes might well be less comely articles than the middle or small sizes is not in much favor to-day, and firms that persist in sending out shoes to the trade that are not as uniform as they can possibly be made will surely lose their hold on even their so called established trade. This matter of greater uniformity is pretty well agreed to and specialization is one of the most effective agencies in attaining it.

The ideal method of readjusting labor costs is to so arrange or rearrage the factory that each operative can earn not only as much but even more money when a reduction in the price of piece-work is made. It makes all the difference in the world just how the work is put up to the workers throughout the factory. This is one of the strong features of specialization, namely, that it allows of such adjustments which are otherwise nearly impossible. The main idea is to simplify the work as much as possible down to the smallest unit, as this promotes productiveness and reduces labor expense. Careful division between skilled and unskilled labor can also be better accomplished in a factory where specialization prevails. There are then strategic advantages to be gained in the physical arrangement of a factory specializing in a certain kind, style and grade of shoes.

A further advantage is to be noted in the purchasing department and in the office. Leather and findings of all sorts can no doubt be even more advantageously bought in a specialized plant. Less liability of errors in the office will also result. The whole tendency will be towards unifying and standardizing the product, and, as a natural result, more uniform methods of making and safeguarding the shoes in the works, of purchasing the leather and findings, of checking the labor costs in the office, and of shipping the goods to the trade will be developed.

One of the chief reasons why the sheet or productive system of moving shoes through the factory has failed to work satisfactorily in many instances is that so many different kinds and styles of shoes are manufactured under the same roof. It is, of course, not at all impossible to work the sheet system successfully under such conditions, but, in factories that specialize in their production, there is no excuse for not running them on the highly productive sheet system. The firm which can supply the trade at the proper time with a large amount of certain kinds of shoes is going to win over a substantial trade because of prompt shipments and uniformity of goods.

Specialization is one of the evolutions that is gradually reshaping the shoe industry, and, as there are so many advantages offered to those who will limit the range of their operations and thoroughly exploit a given field, we may confidently expect that this move in the direction of specialization will continue unimpeded. In fact, everything that tends to promote greater efficiency in the shoe factory must be closely watched and studied. Manufacturers cannot be less efficient than their competitors and hope to reap equal gains from the trade.

CHAPTER VI.
Departments.

I have already alluded to the matter of careful departmental divisions as a factor in the promotion of general factory efficiency. This subject requires special treatment, as a good many manufacturers are unaware of the importance attaching to it.

Some of the most progressive manufacturers have separated the upper leather cutting department from the rest of the factory and also the sole leather cutting as well. By making this a department in itself the leaks that so easily occur may be stopped and the many details may be handled in a more exacting manner than is otherwise possible. It is not difficult to see that if expert men are chosen to handle this part of the business and to devote all their time and energies, money may be saved for the firm. Leather, especially upper stock, is the most expensive and variable element that enters into the building of the shoe and for this reason a firm that is engaged on a fairly large scale may well consider the advisability of establishing a separate upper leather department, if not a sole leather department, entirely distinct from the factory or factories proper. Such a separation must be made absolute or otherwise the valuable double check will be lost.

In the event of a separate upper leather cutting division there is afforded the necessary impetus for the upper leather head or heads to make the best showing possible. Theoretically this might be done without going to the extreme of physically separating the different branches, but in practice it does not work out very successfully. By making a distinct break between the upper cutting and the factory proper the best results have been obtained.

The introduction of the clicking press into the

cutting room has opened up a great many opportunities that did not previously exist and the possibilities of the new order of things have not begun to be realized. With the continued improvements in dies, such as reversible and adjustable dies that allow of different pattern adjustments, the money invested will not need to be so great and the increment on each dollar thus tied up will be greater. Factories which are not large enough to segregate the upper leather cutting into a division by itself will do well to arrange the cutting and fitting departments in such a manner that greater speed and efficiency may result. This is largely a matter of handling the various preliminary operations effectively. Certain operations that are today in many factories being done in the fitting room would much better be done in a department by itself. The skiving, cementing, rolling, marking of uppers and linings can be more effectively done in this room, although usually accomplished in the fitting department. If the factory permits, there are many advantages in having the cutting and fitting departments on the same floor. The work can be more quickly moved from the first to the second department, and, moreover, it will be already on the run when it gets to the machines. There is a great deal in getting the shoes well started in the fitting room, for it is then easier to keep them moving fast until they emerge as fitted uppers.

Evidently there is a field for study in getting the best departmental divisions, that is to say, the most effective division of departments as regards production. Generally speaking, an arrangement which allows of greater facility in producing shoes promotes the quality at the same time. This is not at once apparent, but a little thought will convince one that irregularity in the movement of shoes through a department of the factory reacts on the quality of the shoe. Workmen appreciate the advantage of laboring in a factory where the work comes right and where it comes steadily. Often a rearrangement along the lines suggested will help to make conditions more favorable to smooth and steady production. Naturally, however, each factory must be worked by itself, for what will be a successful line-up and organization of forces in one factory, where such and such conditions have to be met, will not be applicable in another factory where a different set of circumstances have to be considered. It is almost a general axiom, however, that no factory is so well appointed that it could not possibly be bettered in some particular. Often times a mere suggestion or an outside observation leads to the institution of economic changes that greatly promote the general efficiency of the plant. The "good enough" policy is not a safe one to follow for the individual and it is equally unsafe for manufacturing interests. Pickets and outposts still have their valuable uses in the industrial evolutions of the country.

It hardly seems worthy of belief that a firm would arrange or rearrange its departments for greater convenience and accuracy in checking the labor costs, and yet such has often been done. This is unjustifiable, as one will readily see. While it is of the utmost importance that the labor costs shall be accurately and carefully checked, it is totally beyond reason to make the various departments of the factory shape themselves to this end. This is a case of the tail wagging the dog, and in such instances the manufacturer loses sight of the goal altogether. The productive agencies are the ones of first importance, and all other considerations must be subordinated to these producing agents. Along these fundamental lines there are possibilities of attaining greater factory efficiency, and the small no less than the large manufacturer should be keenly alert to institute improvements wherever there is the least reasonable evidence that such a move would better conditions and thus make the plant a more effective producer.

CHAPTER VII.

Shoe Costs

I have tried to emphasize the importance of the fact that each pair of shoes utilizes a certain definite amount of floor space in moving from cutting to the shipping department and this is most valuable information for each manufacturer. The reduction of the protective tariff on finished leather and its products and the whole present-day tendency to close competition makes is highly important, if not indeed imperative, to develop each factory to the highest possible degree of efficiency, and in doing this the exact cost of the shoe must be definitely ascertained.

There are a great many manufacturers who believe that they have their shoe costs figured safely, but I wish to ask have they established their shoe costs on an exact basis? There is a long gap between safe figuring and exact figuring. Not a few manufacturers purposely overfigure their shoe costs, and in so doing believe they are safeguarding their best interests, but it must be asserted, however, that the most farsighted men and the constructive thinkers realize the day of "safe" figuring is rapidly becoming old-fashioned. If for no other reason, the great development and advance of expert accounting that is now going on in all lines of business should spur the careful and progressive shoe manufacturer to demand the most exact records.

In order to reach a sound basis of estimating costs exactly, the manufacturing and labor expense should be carefully seperated from the cost of materials. The reason for this is evident, and yet there are manufacturers who do not seem to grasp it. The fact that better buttons or slightly more expensive fixings are put into some particular shoe and a higher price asked for it, does not imply that the manufacturing expense is greater in proportion to the advanced price placed on the shoe. It is not wise to attempt to take care of future market exigencies in fixing the shoe costs. The manufacturer must be sufficiently farsighted to tell within reasonable limits just what the leather used is going to cost him. Naturally, if the leather is going to advance in price and if he has been unable to foresee it, his whole cost analysis is going to be thrown off. It must be asserted, also, that there are other and more preferable ways of taking care of emergencies which cannot well be foreseen; hence, it is not necessary to over rate the shoe costs to take care of this matter. Overrating costs is no less reprehensible than under rating. Not a few manufacturers have gone to the wall because their profits appeared on paper only. It is possible, both through overfiguring and underfiguring to practically figure one's self out of business.

One of the most common discoveries made in factories which introduce advanced and practical accounting methods is that the profits, which were deemed to accrue from the sale of a particular shoe or shoes, were derived from entirely different sources. Furthermore, it has often been proven that a loss on these "pet" lines has been constantly sustained, a fact of which the manufacturer has been blissfully unconscious. The writer has in mind a certain factory that featured some particularly fancy shoes in its line which were figured at a small profit. It happened, partly through accident, that these shoes were refigured in a different and more exact manner about the middle of the season, and it was proved to the manufacturer's amazement that he was actually losing money on these shoes, and they were consequently dropped from the line. Other instances of a similar nature might be cited illustrating the danger of figuring costs on shoes under or over. It cannot, therefore, be driven home too strongly that there is an exact method for every manufacturer to establish his shoe costs, but it is not within the province of such a discussion as this to go into specific details, nor would it be wise to do so.

In the light of what has been stated above, in order to get at a sure and exact cost basis, the different elements, such as materials, labor and manufacturing expense, selling expense, etc., must be carefully seperated and sub-divided. A system which figures the cost flatly on a pair of shoes cannot possibly be exact. It is not reasonable that a shoe made to sell for $1.50 should bear a proportion of the manufacturing expense equal to that borne by a shoe made to sell for $2.50 or more. Unless the proper proportions are ascertained and the subdivision of burdens is carefully analyzed, the shoe costs cannot possibly approach the degree of exactness which the trade conditions of to-day are demanding more and more. It should further be admitted that good salesmanship will cover a multitude of sins in the cost department, but the great shoe manufacturing industry cannot be conducted on a hit-or-miss principle nor depend on its salesmen to overcome through sheer ability the effects of rudimental errors in the shoe costing system. Possibly many firms never realize how frequently their salesmen overcome serious difficulties of this nature.

Factory conditions are many and varied; in fact their name is legion. The best one can hope to do in a general treatment is to give a clear line of thought on these important matters and allow the reader to fill in the outlines. Exact systematic book records and improved accountancy methods are doing a great deal toward perfecting the great shoe industry of the country. Lax methods of making and safeguarding the product in the works are being effectually eliminated. The evolution is constantly going on, and it is felt in every root and fibre of the business.

CHAPTER VIII.

Conclusion

In summing up this discussion we are led to believe that the shoe business, as now conducted, is far too lax in methods of production. Compared with other great industries this fact is strongly emphasized. It is fairly well agreed by all, that the shoe business is one of the most strongly individualized that we know of especially when we consider the magnitude of the enterprise. The main trend is going to be towards standardization and solidification. This will be accomplished by methods and systems that sprout and grow from the mother-stem of practical efficiency. Those who pooh-pooh this idea of system as a ball of red tape are only blowing sand into their own eyes. The experimental stage has long been passed and the alert manufacturer must discriminate between his apparent and real needs; this means that the Sham or red tape must no longer be misconstrued with the Real, or advanced productive methods.

All instances of irregularity in records or lack of records are now investigated with an eye to obtaining more exact data. Some firms keep a seperate book for lasts, in which are recorded the principal measurements, such as heel spring, toe spring, extention, waist, ball, etc., and by reference to this book the manufacturer can tell at a glance whether or not a new style of last will suit his trade. Naturally the good fitting lasts have certain cardinal measurements which are practically always the same and the good fitters are usually the good sellers. This book is in a way a specific record of what his trade demands, and any new effects that he may wish to add to his line of shoes must first meet these necessary requirements which are known and carefully catalogued. This sort of a record saves the manufacturer a great many dollars and prevents him from loading up on a last that his trade as a whole will not use. Another set of records keeps the manufacturer informed accurately as to the amount of floor space used in moving a pair of shoes. These two sets of records indicate the energy with which some shoe manufacturers are analyzing their business, and they are treated in conjunction to show that statistics, other than those pertaining directly to the profit and loss part of the business, are distinctly worth while and cannot wisely be neglected.

The sheet or productive system, the great tendency towards specialization, and the careful arrangement of the departmental divisions, are all factors of great importance in attaining greater factory efficiency. These do not in any sense comprise all the factors that make for the desired end. It would be a pretty difficult task to try to mention all the productive efficiency factors and give them their proper proportion to the whole. The task would be further complicated by the fact that what is good for one is not necessarily good for another, where two distinct sets of conditions, geographical or inherent, have to be considered. This limits one to a more or less general discussion, but does not prevent one from drawing some important inferences which a clear application of thought will turn to advantage.

The trend of events indicates that the shoe industry is passing through an important evolution, and, in the light of what has and is being done, it would appear that it is being reclaimed gradually but by sure degrees from a chaotic state. From the old-time hit-or-miss, hap-hazard, come-a-day-go-a-day methods of manufacturing shoes to the present-day system of production is indeed a long leap. The movement, however, is still on and there is a vast field ahead. Each new system that works successfully does its part to help advance the industry. In the light then of what has already been accomplished one cannot help but believe in the possibilities that present themselves. The small factory finds it harder to make money than formerly and the most efficient small concerns soon become large ones. This tendency will continue and the size of the plant will in a measure be a proof of its productive efficiency. As the manufacturer's profit diminishes he must either make more shoes than formerly or increase the efficiency of his plant to such an extent that the cost of production will be less. This is a point that the manufacturer who reads these discussions should not fail to grasp. Waste of materials is not the only kind of waste that the shoe manufacturer must guard against. Lack of efficiency is a waste that mounts into dollars much faster than a great many manufacturers realize.

www.ingramcontent.com/pod-product-compliance
Lightning Source LLC
LaVergne TN
LVHW041520070426
835507LV00012B/1701